TABLE OF CONTENTS

SUMMARY iii

1 INTRODUCTION 1

2 NANOTECHNOLOGY BACKGROUND 3

3 APPROACH 5

4 OPPORTUNITIES FOR REDUCING FRESHWATER CONSUMPTION IN 6
  COAL-FIRED POWER PLANTS 6

   4.1 Use of Nontraditional Waters 6
   4.2 CO_2 capture 6
   4.3 Removal of NO_x and SO_2 6
      4.3.1 NO_x Removal 7
      4.3.2. SO_2 Removal 7
   4.4 Coolants 8
   4.5 Plant Efficiency 8
      4.5.1 Air Heater Inefficiencies 8
      4.5.2 Corrosion in Boilers 8
      4.5.3 Higher-Temperature Operations 9
      4.5.4 Fuel Efficiency 9
      4.5.5 Strain on Older Components 9
   4.6 Storage 9

5 NANOTECHNOLOGY APPLICATIONS 11

   5.1 Membranes 11
      5.1.1 Water Purification 11
      5.1.2 Membranes for CO_2 Capture 17
   5.2 Coatings and Lubricants 18
      5.2.1 Coatings to Insulate and Reduce Heat Loss Throughout the Plant 18
      5.2.2 Coatings to Limit/Prevent Corrosion in Utility Boilers 19
      5.2.3 Coatings to Improve Corrosion, Erosion, and Wear Resistance 20
      5.2.4 Coatings to Improve Reliability and Fuel Efficiency 20
      5.2.5 Lubricants with High Chemical and Physical Stability 21
      5.2.6 Welding Materials 21
   5.3 Catalysts and Enzymes 22
      5.3.1 Enzymes to Improve Fuel Efficiency 22

  5.3.2 Catalysts to Improve Efficiency of Sulfur Removal 22
 5.4 Materials that Can Withstand High Temperatures 23
 5.5 Nanofluids 23
 5.6 Nanosensors 25
  5.6.1 Strain and Impact Damage Identification 26
  5.6.2 Water Quality Detection and Monitoring 27
  5.6.3 Measuring Mercury in Flue Gas 27
 5.7 Batteries and Capacitors 28

6 CONCLUSIONS AND RECOMMENDATIONS 30

7 REFERENCES 32

SUMMARY

This report was funded by the U.S. Department of Energy's (DOE's) National Energy Technology Laboratory (NETL) Existing Plants Research Program, which has an energy-water research effort that focuses on water use at power plants. This study complements the overall research effort of the Existing Plants Research Program by evaluating water issues that could impact power plants.

A growing challenge to the economic production of electricity from coal-fired power plants is the demand for freshwater, particularly in light of the projected trends for increasing demands and decreasing supplies of freshwater. Nanotechnology uses the unique chemical, physical, and biological properties that are associated with materials at the nanoscale to create and use materials, devices, and systems with new functions and properties. It is possible that nanotechnology may open the door to a variety of potentially interesting ways to reduce freshwater consumption at power plants.

This report provides an overview of how applications of nanotechnology could potentially help reduce freshwater use at coal-fired power plants. It was developed by (1) identifying areas within a coal-fired power plant's operations where freshwater use occurs and could possibly be reduced, (2) conducting a literature review to identify potential applications of nanotechnology for facilitating such reductions, and (3) collecting additional information on potential applications from researchers and companies to clarify or expand on information obtained from the literature.

Opportunities, areas, and processes for reducing freshwater use in coal-fired power plants considered in this report include the use of nontraditional waters in process and cooling water systems, carbon capture alternatives, more efficient processes for removing sulfur dioxide and nitrogen oxides, coolants that have higher thermal conductivities than water alone, energy storage options, and a variety of plant inefficiencies, which, if improved, would reduce energy use and concomitant water consumption. These inefficiencies include air heater inefficiencies, boiler corrosion, low operating temperatures, fuel inefficiencies, and older components that are subject to strain and failure.

A variety of nanotechnology applications that could potentially be used to reduce the amount of freshwater consumed — either directly or indirectly — by these areas and activities was identified. These applications include membranes that use nanotechnology or contain nanomaterials for improved water purification and carbon capture; nano-based coatings and lubricants to insulate and reduce heat loss, inhibit corrosion, and improve fuel efficiency; nano-based catalysts and enzymes that improve fuel efficiency and improve sulfur removal efficiency; nanomaterials that can withstand high temperatures; nanofluids that have better heat transfer characteristics than water; nanosensors that can help identify strain and impact damage, detect and monitor water quality parameters, and measure mercury in flue gas; and batteries and capacitors that use nanotechnology to enable utility-scale storage.

Most of these potential applications are in the research stage, and few have been deployed at coal-fired power plants. Moving from research to deployment in today's economic environment will be facilitated with federal support. Additional support for research development and deployment (RD&D) for some subset of these applications could lead to reductions in water consumption and could provide lessons learned that could be applied to future efforts. To take advantage of this situation, it is recommended that NETL pursue funding for further research, development, or deployment for one or more of the potential applications identified in this report.

1 INTRODUCTION

This report was funded by the U.S. Department of Energy's (DOE's) National Energy Technology Laboratory (NETL) Existing Plants Research Program, which has an energy-water research effort that focuses on water use at power plants. This study complements the overall research effort of the Existing Plants Research Program by evaluating water issues that could impact power plants.

A growing challenge to the economic production of electricity from coal-fired power plants is the demand for freshwater, particularly in light of the projected trends for increasing demands and decreasing supplies of freshwater. NETL's Existing Plants program has a number of research activities underway to help meet these challenges, including targeted research applications such as specific advanced cooling technologies, the use of nontraditional water sources, and water reuse and recovery. In addition to conducting and sponsoring detailed investigations into such areas, NETL also recognizes the importance of exploring new approaches for reducing freshwater use.

One new approach comes from the field of nanotechnology. Nanotechnology uses the unique chemical, physical, and biological properties that are associated with materials at the nanoscale (a nanometer is one billionth of a meter) to create and use materials, devices, and systems with new functions and properties. It is possible — given the development of nanomaterials applications in other fields such as medicine, electronics, computer science, transportation, and agriculture — that nanotechnology may also open the door to a variety of potentially interesting ways to reduce freshwater consumption at power plants. Today, there are relatively few commercial applications of nanotechnology to reduce water consumption at coal-fired power plants. Indeed, most such applications have yet to be identified. It is important, therefore, at this time, to begin considering how applications from this emerging field could help address the water use issue for power plants so that decision makers can consider these options for potential research development and deployment (RD&D) support. An early view of these potential applications can provide NETL with information to explore these ideas and to identify promising nano areas for RD&D support. Given the range in the status of development of nano options for power plant use, such support could range from expanding research in specific areas, to supporting the conduct of pilot test of promising nanotechnology applications in power plants, to promoting the deployment of technologies that have shown success in other industries to power plants.

According to the National Nanotechnology Initiative (NNI), the program that coordinates federal nanotechnology research and development, nanotechnology has the potential to profoundly change our economy and to improve our standard of living, in a manner not unlike the impact made by advances over the past two decades by information technology. While some commercial products are beginning to come to market, many major applications for nanotechnology are still five to ten years out. (Source: http://www.nano.gov/html/facts/faqs.html.)

This report provides an early overview of how applications of nanotechnology could potentially help reduce freshwater use at coal-fired power plants. It was developed by identifying areas within a coal-fired power plant's operations where freshwater use occurs and could possibly be reduced, conducting a literature review to identify potential nano applications for facilitating such reductions, and collecting additional information on potential applications by contacting researchers and companies and/or reviewing company website information for further explanation and understanding of how such technologies may be used. It does not address potential environmental, safety, and health (ES&H) issues, nor does it address costs. ES&H issues are always a concern with any new technology, and research is underway to identify and analyze potential impacts of specific types of nanoparticles, but attempting to link potential ES&H impacts with the technology applications described in this overview is beyond the scope of this overview. Also, because many, if not most, of the potential applications identified in this study are not yet commercial, very little cost information exists.

The remainder of this report consists of five chapters. Chapter 2 provides a brief background on nanotechnology. Chapter 3 describes the approach used to identify potential applications for reducing water consumption at power plants. Chapter 4 summarizes areas within coal-fired power plants where freshwater is used and has the potential for reduction. Chapter 5 presents various nanotechnology-based approaches for reducing freshwater use in the areas identified in Chapter 4, and Chapter 6 presents some conclusions and recommendations regarding next steps.

2 NANOTECHNOLOGY BACKGROUND

While there is no universally accepted definition of nanotechnology, it is generally understood to involve the manipulation of matter on a near-atomic scale to produce new structures, materials, systems, catalysts, and devices that exhibit novel phenomena and properties. It is a growing research area that may yield far-reaching changes in virtually everything that affects our lives.[1]

At the nanoscale level (a nanometer is one billionth of a meter), materials exhibit unique properties that affect their physical, chemical, and biological behaviors. For example, because nanoscale materials (nanoscale materials range in size from about 1 to 100 nanometers) have very large surface areas relative to their comparable bulk materials, they have more surfaces available for interactions with other materials around them. Nanomaterials can display characteristics such as high thermal and electrical conductivity, extraordinary strength, and enhanced magnetic properties, which make them attractive for a variety of thermal, electrical, magnetic, catalytic, electronic, and other applications. For example, the carbon nanotube is a one-atom thick cylinder of graphite that is about six times lighter than steel, but can support weight that is 40 times heavier. Carbon nanotubes can be created in various shapes, which can in turn affect their properties. A single-walled carbon nanotube (SWNT), for example, can have a surface area equal to 1,600m^2/gram; that means that 4 grams of nanotubes has the area of a football field (Shelley 2008). A substance that is composed of nanomaterials has many more points of contact where reactions can occur than the same quantity of that substance at conventional size. Existing industrial applications include industrial measurement and sensing devices; thin films that are water-repellent, electrically conductive, self-cleaning, anti-microbial, or scratch-resistant; improved catalysts; microelectronics; and others. Consumer applications include wrinkle-free clothing and transparent composite materials such as sunscreens containing nanosized titanium dioxide particles, and tennis racquets that use carbon nanotubes to stiffen key areas of the racquet head and shaft. The Project on Emerging Nanotechnologies (http://www.nanotechproject.org/) has prepared inventories of hundreds of existing nanotechnology applications. Research is underway on applications in the fields of energy, agriculture, transportation, medicine, computing, electronics, and manufacturing. Applications in various stages of research and development include filters and other techniques for water purification, efficient insulation materials, new batteries and other storage devices, low-cost methods to detect impurities; new drug-delivery techniques with fewer side effects, and many more.

DOE has suggested how nanotechnology may impact energy in the future: "Understanding the emergent behavior of materials and chemical reactivity at the nanoscale

---

[1] The global nanotechnology market is projected to grow at a Compound Annual Growth Rate of over 18 percent during 2010–2013; the global market for nanotechnology incorporated in manufactured goods is projected to be worth US$ 1.6 trillion, representing a compound annual growth rate of around 50 percent over the 2101–2013 forecast period 2010–2013). Massive government and corporate investments in nanotechnology R&D is expected to drive this prospective growth.
(http://www.marketresearch.com/product/display.asp?productid=2621511)

offers remarkable opportunities in a broad arena of applications including solid-state lighting, electrical generators, clean and efficient combustion of 21st-Century transportation fuels, catalytic processes for efficient production and utilization of chemical fuels, and superconductivity for resistance-less electricity transmission." (Source: *Energy Frontier Research Centers*, available at http://www.er.doe.gov/bes/brochures/files/EFRC_brochure.pdf.) While much of this potential is yet to be developed, a significant amount of research and development (R&D) regarding nanotechnology has occurred, and much is underway that applies directly to the energy-water links at coal-fired power plants.

3 APPROACH

The following three-step approach was used to identify potential nanotechnology applications for reducing freshwater consumption:

1. Identified aspects of coal-fired power plant operations where the amount of freshwater consumption could be reduced.

2. Conducted a literature review to identify possible ways in which nanotechnology or nanomaterials could be used to reduce water consumption in those areas.

3. Obtained additional information on potential applications by contacting researchers and companies and/or reviewing company website information to clarify or verify information obtained in the literature search.

It is important to note that the literature contains very few applications of nanotechnology that are specifically directed toward reducing water consumption — and fewer still that address water consumption in coal-fired power plants. In many cases, potential applications for reducing water consumption in coal-fired power plants had to be inferred from reports on their use in related applications or from early basic research findings.

Also, the level of response obtained from further inquiries (Step 3) varied from no response to more than 100 pages of information including a proposal for a pilot study at a power plant. The findings presented in Chapter 5 reflect this variability in application status and detail of available information. Some potential applications are much further along the development path than others, and as a result, the descriptions of the various potential applications vary in detail and length. Because the objective of this task is to provide an overview of potential applications, we included several applications where data and information were limited, so that the reader would have a feel for the breadth and depth of these applications (rather than omitting them simply because detailed information was not available). The resources for the current task did not allow for a detailed investigation of all of the applications for which information was limited. However, subsequent, in-depth investigations can be conducted on any of applications presented in this report should NETL agree that further information would be useful for directing RD&D efforts.

4 OPPORTUNITIES FOR REDUCING FRESHWATER CONSUMPTION IN COAL-FIRED POWER PLANTS

This chapter identifies opportunities for and means of reducing freshwater consumption in coal-fired power plants. It includes opportunities such as using nontraditional waters, reducing the water required to capture carbon dioxide (CO_2) and to remove nitrogen oxides (NO_x) and sulfur dioxide (SO_2), and providing better coolants. It also highlights areas in power plants where efficiency could be improved, because increased efficiency means less energy consumption, and — because energy production requires water — less water consumption.

4.1 USE OF NONTRADITIONAL WATERS

NETL has been investigating the use of nontraditional waters including brackish and saline water supplies and various domestic and industrial wastewaters in coal-fired power plants as a means of reducing freshwater use. A critical concern regarding the use of nontraditional water is the treatment required to improve the quality for use process and cooling water systems. Any improvements in the efficiency of the technologies and systems required to treat these waters will increase the ability to use them, and hence, reduce freshwater use. Common contaminants that need to be removed or reduced prior to use include ammonia, cadmium, zinc, and other metals; chloride, chlorine, dissolved solids, nitrogen, organic compounds, phosphates and phosphorous-containing compounds; and silica, which can contribute to biofouling of membranes, corrosion, and scale.

4.2 CO_2 CAPTURE

The most common post-combustion technology for capturing CO_2 in use today is a chemical absorption process that uses amines. In this process, the exhaust gases flow through several baths in which the carbon dioxide is bound with amines (typically monoethanolamine (MEA). The MEA is then pumped to a regenerator, where steam is used to separate the CO_2. Contaminants are removed from the MEA, which is then recirculated back to the absorber. This approach can remove 90–95 percent of the CO_2 from the flue gas, but it requires diversion of steam from the generating process and a significant amount of the energy (and with that the associated water) to compress the CO_2 once it is captured.

4.3 REMOVAL OF NO_X AND SO_2

The technologies used to remove regulated pollutants from flue gases use energy and water. By increasing the efficiencies of these technologies, less energy, and consequently, less water, will be used. Related to the increased efficiency of pollutant removal is the detection and monitoring of pollutants to ensure that the technologies are removing the required amounts. Utilities can be fined for failing to remove enough of the pollutants, and/or they may need to invest in new equipment or process changes. On the other hand, by removing more than what is required, unnecessary amounts of energy (and water) may be consumed. New regulations on emissions can be expected to increase the need for monitoring of more sources on larger scales and for more constituents.

4.3.1 NO_x Removal

NO_x removal from coal-fired power plants is typically accomplished by the Selective Catalytic Reduction (SCR) process. The basic principle of SCR is the reduction of NO_x to elementary nitrogen (N2) and water by the reaction of NO_x and ammonia (NH_3) within a catalyst bed. The SCR process is effective in removing NO_x, but it also reacts with the sulfur in the coal to produce a byproduct, ammonium bisulfate. Ammonium bisulfate has a significant detrimental effect on the efficiency and reliability of the air regenerative heater(s) in the power plant.

4.3.2. SO_2 Removal

Over the next several years, increased sulfur removal will be required to (1) meet new regulatory requirements and (2) reduce the sulfur content to allow for efficient CO_2 capture. On June 22, 2010, the Environmental Protection Agency (EPA) revised its primary National Ambient Air Quality Standard (NAAQS) for SO_2 (EPA 2010). The new rule replaces its two existing primary standards (140 parts per billion [ppb] over a 24-hour period and 30 ppb over a year), which did not account for short-term peaks in emissions, with a new 1-hour standard of 75 parts per billion (ppb). According to the new rule, states with areas that do not meet the new standard must submit plans by 2017 showing how they will come into compliance – generally by requiring the sources of the SO_2 emissions (e.g., power plants) to upgrade capture equipment.

The relationship between CO_2 capture and sulfur removal may be an even greater driver for more efficient sulfur removal technologies than the regulatory requirement. This is because the likely near-term possibilities for CO_2 capture (post-combustion solvent scrubbing of flue gas) require that nearly all of the SO_2 and NO_x be removed from the flue gas prior to CO_2 capture to slow the degeneration of the CO_2 reagent (Smith et al. 2008). Power plants with existing flue gas desulfurization (FGD) systems will need to modify their processes to achieve ultra-high SO_2 removal, and plants that do not already have FGD will need to consider installing systems that achieve up to 99.8 percent SO_2 removal efficiency (Smith et al. 2008). Wet scrubbing with lime or limestone slurries will likely be the most reliable means for achieving the high levels of SO_2 removal required to accommodate CO_2 capture using a chemical absorption process (Smith et al. 2008). This approach is currently the dominant commercial FGD technology; worldwide, about 80 percent of coal-fired power plants, representing about 87 percent of capacity, use FGD systems that are based on lime or limestone wet scrubbing (Nolan 2000). While they have relatively high removal efficiencies, wet FGD systems consume a lot of water. In plants that recirculate cooling water and use wet FGD systems, 10 percent of the evaporative cooling water loss is in the FGD system (with 90 percent in the cooling towers) (NETL 2006). Clearly, FGD systems that consume less water are needed.

4.4 COOLANTS

Cooling is a key issue for coal-fired power plants, and it will continue to be so as heat generation rates continue to increase and as regulations regarding cooling methods (i.e., the Clean Water Act's Section 316(b) requirements that effectively limit the use of efficient once-through cooling systems) are implemented. The conventional heat transfer fluid at power plants — water — has an inherently low thermal conductivity relative to solids. New coolants that could encompass the thermal characteristics of high-conductivity solids could increase efficiency and thereby reduce water use.

4.5 PLANT EFFICIENCY

Because energy production requires water, improvements in power plant efficiency will also reduce water use. For example, in plants with recirculating cooling systems, evaporation loss is proportional to the cooling load of the tower (Kim and Smith 2004). Therefore, makeup water can be reduced by decreasing the cooling load. Cooling load can be reduced by improving the energy efficiency of various processes within the plant. Examples of areas within the power plant where efficiencies can be improved include the following:

4.5.1 Air Heater Inefficiencies

Air-heater inefficiencies are caused by corrosion and fouling associated with burning coal and operating NO_x pollution control devices. As noted earlier, a side effect of the SCR process to reduce NO_x is the production of ammonium bisulfate, which causes heavy plugging, corrosion, and efficiency reductions in the plant's air heaters, and in turn, means increased fuel consumption, air heater repairs, and lost production because the plant does not operate while repairs are underway. Historically, there has been no effective way to eliminate or reduce the air heater issues associated with ammonium bisulfate deposits.

4.5.2 Corrosion in Boilers

The Electric Power Research Institute (EPRI) reports the following concern with corrosion of boiler waterwalls (EPRI 2008): Fireside corrosion of boiler waterwalls continues to be the number one issue resulting in forced outages and boiler unavailability for conventional coal-fired fossil power plants. Although several types of coatings and weld overlays have been used to extend the service life of boiler tubes, these coatings and weld overlays only provide protection for 1 to 8 years in subcritical boilers. Because ultra supercritical boilers operate at much higher temperatures and pressures than subcritical or supercritical boilers, accelerated and severe fireside corrosion in ultra supercritical boiler waterwalls is anticipated to be a primary concern.

There is a need for improved coatings or claddings to reduce/eliminate waterwall damage by mitigating fireside corrosion. There is also a need for technologies to identify potential corrosion problems so that they can be addressed before they lead to failures and outages. Current boiler operational practice is to remove samples for determination of deposit loading and the need for chemical cleaning of the boiler. Because sample removal locations are often determined on the basis of experience, decisions are based on incomplete and or erroneous information (Intertek 2010).

4.5.3 Higher-Temperature Operations

The thermodynamic efficiency of the steam cycle in the power plant increases with increasing temperature, but today's plants are only about 35 percent efficient. One of the barriers to operating at higher temperatures is that many of the metals and other materials currently used in power plants cannot withstand high temperatures without degradation or oxidation. The use of materials that can withstand these high temperatures will help allow the plant to operate at higher temperatures and thereby, more efficiently. According to Tomar (2008), the use of better and improved high-temperature structural materials, the power generation efficiency of the power plants can be increased by 15 percent.

4.5.4 Fuel Efficiency

The characteristics of the coal burned in the power plant help determine the combustion efficiency and the types and amounts of resulting pollutants. If the combustion efficiency of the coal can be increased, less coal can be used, and the overall efficiency of the power plant can be increased. Similarly, reductions in the amounts of pollutants produced during combustion will also increase plant efficiency.

4.5.5 Strain on Older Components

EPRI (2010) reports that the majority of fossil-fueled plants worldwide are more than 30 years old, and that increased demands for operational flexibility put additional strain on existing components and materials such as weld joints. Weld failures can contribute to forced shutdowns (Saha et al. 2010) and attendant operational inefficiencies. Technologies that can reduce weld failures will help improve plant efficiencies.

4.6 STORAGE

Power plants operate more efficiently when run at a constant optimized power output, rather than at variable outputs, which occur in response to short-term variations in power demand. Operating at a variable output is inefficient because it puts additional strain on the machinery (which can accelerate wear and require replacement of components sooner than if operated at a constant output) and because it consumes more fuel per given unit of power output. Short-term demands are likely to increase in the future as the intermittent contributions of wind and solar power enter the grid, and as existing sources are called upon to increase or decrease their contributions to smooth the load. Even though coal-fired plants are generally considered baseload plants that produce power at a constant rate, newer coal-fired plants can accommodate load changes, and improved designs allow power to be provided in response to changing needs of the grid. Techniques that would allow coal-fired power plants to store electricity generated that is not needed during periods of low demand and to release it later when demand increases could help power plants operate at a more constant, efficient rate. To date, utility-scale storage has been primarily pumped hydro (where water is pumped uphill when power is cheap and then released to spin a turbine during peak demand periods) and compressed air energy storage (where off-peak power is taken from the grid and used to pump air into a sealed underground cavern to a high pressure, which is then kept underground for periods of peak use). Until recently, demand has been fairly predictable, and there was little incentive to develop new ways to add storage beyond those provided by pumped storage, especially when it is questionable that the regulatory structure would allow utilities to develop an unproven technology such as utility-scale storage via advanced batteries or capacitors. However, as grid demands are projected to become less predictable as solar and wind power are integrated, the need for additional storage capacity will increase. A bill has been introduced in the California legislature (Assembly Bill 2514) that would require the state's utilities to match 2.25 percent of their peak loads with energy storage by 2014 and 5 percent of their peak loads by 2020.

5 NANOTECHNOLOGY APPLICATIONS

This chapter identifies several potential nanotechnology applications — in various phases of development — that can be considered for use in reducing freshwater consumption at coal-fired power plants. These potential applications can be used either directly (e.g., nanofiltration to purify water allowing greater use of nontraditional sources) or indirectly (e.g., through improved plant efficiency, which lowers energy use and hence, water consumption.) Potential nanotechnology-based applications include membranes; coatings and lubricants; catalysts and enzymes;, materials that can withstand high temperatures; coolants; sensors; and batteries and capacitors.

5.1 MEMBRANES

Nanotechnology is being used to improve membranes both for water purification (e.g., nanofiltration, reverse osmosis) and for CO_2 capture.

5.1.1 Water Purification

Membranes and sorbents that use nanotechnology or contain nanomaterials can help improve the quality of water prior to use in boilers or cooling systems. This can help facilitate the use of nontraditional waters such as municipal treated water/reclaimed water and mine pool waters, as well as water recycled from cooling water recirculating systems.

Membrane technologies include microfiltration, ultrafiltration, nanofiltration, and reverse osmosis (RO). In general, microfiltration is used to reduce turbidity and solids (particles of 50 nm or larger), ultrafiltration to further reduce turbidity and remove high molecular weight dissolved substances (particles of about 3 nanometers [nm] or larger), and nanofiltration to remove hardness (e.g., multivalent cations), dissolved constituents, natural organic matter (NOM), biological contaminants, organic pollutants, and nitrates (particles about 1 nm or larger). With RO, pressurized water passes through a semi-permeable membrane with pore sizes of 0.1–1.0 nm to separate dissolved substances including salts and organic substances. Often one or more of these membrane technologies will be used in an integrated sequence to gradually remove impurities and limit membrane fouling in the downstream membranes.

Table 1. Characteristics of Four Membrane Filtration Technologies

Parameter	Microfiltration	Ultrafiltration	Nanofiltration	Reverse Osmosis
Pore Size	0.01–1.0 µm	0.001–0.01 µm	0.0001–0.001 µm	<0.0001 µm
Molecular Weight Cutoff	>100,000	1,000–300,000	300–1,000	100–300
Operating Pressure	<30 psi	20–100 psi	50–300 psi	225–1,000 psi
Membrane Materials	Ceramics, polypropylene, polysulfone, polyvinylidene-difluoride	Ceramics, polysulfone, polyvinylidenedi-fluoride, cellulose acetate, thin film composite	Cellulose acetate, thin film composite	Cellulose acetate, thin film composite, polysulfonated polysulfone
Membrane Configuration	Tubular, hollow-fiber	Tubular, hollow-fiber, spiral-wound, plate and frame	Tubular, spiral-wound, plate and frame	Tubular, spiral-wound, plate and frame
Types of Materials Removed	Clay, bacteria, viruses, suspended solids	Proteins, starch, viruses, colloid silica, organics, dyes, fats, paint, solids	Starch, sugar, pesticides, herbicides, divalent anions, organics, BOD, COD, detergents	Metal cations, acids, sugars, aqueous salts, amino acids, monovalent salts, BOD, COD

Abbreviations:

psi = pounds per square inch; µm = micrometer; BOD = biochemical oxygen demand; COD = chemical oxygen demand

Key factors affecting the performance of a membrane include its (1) permeability (the volume of water flowing through the membrane per unit of area and time) and (2) selectivity, or separation factor (the difference in permeability of the components of interest). Nanotechnology has contributed to recent developments in the material science of membranes. Permeability, selectivity, and resistance to fouling have been improved by using newly available nanomaterials. The high surface areas and throughputs of nanomaterials are being used and continue to be investigated as ways to improve membrane technologies. Also, the catalytic properties of some nanomaterials have the potential to neutralize chemicals and microorganisms. A variety of nanomembranes with varying characteristics are available, and selection for a given set of conditions would be made on the basis of the desired product water and the quality of the incoming feed water. Nanotechnology can also be used to advance RO membrane performance. As noted, bacterial cell adhesion or fouling tends to increase energy consumption and decreased sustainable fluxes through the membrane. By incorporating nanomaterials into RO membranes, performance can be improved, by controlling membrane roughness, hydrophilicity, and surface charge, thereby inhibiting the adhesion of bacteria cells and consequent biofilm coverage.

Sorbents are used as separation media in water purification techniques to remove organic and inorganic pollutants. Nanoparticles can be attractive sorbents for at least two reasons. First, as noted earlier, they have much larger surface areas than their bulk counterparts. Second, their affinity toward target compounds can be increased when they are functionalized with various chemical groups. Research has found that metal-ion sorption capacities of multiwalled carbon nanotubes, for example, were three to four times larger than those of activated carbon (a commonly used sorbent in water treatment) (Savage and Diallo 2005).

Existing NETL-funded Nanofiltration Research

Sandia National Laboratories (SNL) is conducting a study for NETL to evaluate nanofiltration (NF) treatment options to enable use of two types of nontraditional water sources (cooling tower recirculating water and produced water from oil and gas extraction) as alternatives to freshwater make-up for thermoelectric power plants. Both of these sources contain moderate to high levels of total dissolved solids (TDS). While RO is the most mature and commonly considered option for treating water with high TDS levels, it is generally too expensive to be a feasible option for treating produced waters for power plant use. Therefore, SNL is investigating the use of NF as a potentially more cost-effective treatment option. Although NF is not as effective as RO for the removal of TDS (the typical salt rejection rate for NF is about 85 percent, while that for RO is greater than 95 percent), NETL (2010) notes that the NF performance should be sufficient for typical power plant applications. An NF system is expected to have lower capital costs than an RO system, and, because it requires less pressure to achieve an equivalent flux of product water, it is also expected to have lower operating costs. The study is projected to be completed in 2010 (NETL 2008).

Other NF and Membrane-related Nanotechnology Research and Applications

A variety of other membrane RD&D activities in various stages of development are underway. The following examples are arranged roughly according to level of development, with those more in the research stages followed by those closer to or at the deployment stage.

- *Molecular manipulation of nanofilters.* The potential for further advancements in nanofiltration includes research on the manipulation of nanofilters on the molecular level. This could lead to the further development of surfaces that reduce fouling. Filters that would select molecules on the basis of attributes other than size would allow NF to remove ions at the level removed by RO, but at lower cost.

- *Incorporating nanoparticles with specific functions.* Silver nanoparticles have been incorporated into the polymer matrix of a membrane to inhibit biofouling. The antibacterial capacity associated with the gradual release of ionic silver by the prepared nanocomposites can be effective in reducing intrapore biofouling in nanocomposite membranes of a wide range of porosities. Such nanocomposites can also be used as materials to inhibit the biofilm growth on downstream membrane surfaces. (Source: *PERMEANT: Partnership for Education and Research in Membrane Nanotechnologies*, available at http://www.egr.msu.edu/permeant/index.html.)

- *Membranes to reduce mineral scaling.* Researchers at the University of California at Los Angeles recently reported results of their investigations into a new class of nanostructured membranes that have resulted in a significantly lower propensity for mineral scaling compared to existing commercial membranes with the same salt rejection rates (about 95 percent) and surface roughness (about 70 nm). See Lin et al. (2010) for more information on these new membranes.

- *Nanoparticles that increase the flow of water.* With conventional RO membranes, high pressure is needed to force the molecules through the membranes. Hydrophilic nanoparticles allow the water molecules to pass more easily through the membrane, thereby lowering the amount of pressure needed to separate the water from the salt, and hence, the energy requirements.

- *Nano-enhanced photocatalysts.* The following information comes from Li et al. (2010) and Wu (2010). Researchers at the University of Illinois have been using nanoparticles in a new process that uses sunlight or artificial light to (1) continue the disinfection process after the light source has been removed and (2) increase disinfection rates. Irradiating water with high-intensity ultraviolet light kills bacteria. The researchers have been working to enhance the effectiveness of this method by adding a photocatalyst that is activated by visible (not ultraviolet [UV]) light and generates reactive chemical compounds that break down microbes into CO_2 and water. The photocatalyst consists of fibers of titanium oxide doped with nitrogen to make it absorb visible light, and nanoparticles of palladium to increase the efficiency of the disinfection. In their experiments, after one hour of shining a halogen desk lamp on the solution to be disinfected, the concentration of bacteria dropped from 10 million cells per liter to 1 cell per 10,000 liters). Further, after removing the light source, the photocatalyst remained active — after a few minutes of exposure to the light source, the photocatalyst remained active for up to 24 hours. (For most catalysts, activity stops once the light source is removed.) The researchers explain that the palladium nanoparticles increase the power of the photocatalyst in two ways. First, when photons hit the material, they create pairs of positive charges or holes, and negative charges (electrons). The positively charged holes react with water to produce hydroxyl radicals that attack the bacteria. The nanoparticles move the electrons away from the holes to increase efficiency, and in the process, the nanoparticles enter a different chemical state and store the negative charges. When the light is removed, the charge is slowly released, providing the memory effect that allows the catalyst to continue working. Thus, the photocatalyst can disinfect at full power during the day, or when a light source is on, and then continue to work at night, or when the light source is removed. The researchers suggest that because the disinfection process is quick, systems could be designed to clean large volumes of water by exposing it to light as it flows through the pipe.

- *Hybrid systems.* Nanotechnology can be used with other technologies to improve process efficiency. For example, researchers at Michigan State University have combined nanoparticle-based functionalities with membrane separation. In this hybrid system, ozonation (a process for infusing water with ozone to kill bacteria and other microorganisms) mitigates membrane fouling, because foulants are oxidized by ozone and/or hydroxyl radicals. The introduction of nanoparticles such as iron oxide (Fe_2O_3) and manganese dioxide (MnO_2) at the membrane surface can significantly enhance the efficiency of the process. This enhancement is because of the catalytic effect of the nanoparticles and more targeted oxidation of the NOM portion that is concentrated at or near the membrane surface contributing to membrane fouling. Because the surface remains relatively foulant-free, the operational life of the membrane can be increased. (Source: *Self-cleaning Ceramic Membranes for Removal of Natural and Synthetic Nanomaterials from Drinking Water Using Hybrid-Ozonation Membrane Filtration*, available at http://www.egr.msu.edu/nirt/membrane/.)

- *Membranes with carbon nanotubes.* A process to manufacture microfabricated membranes in which aligned carbon nanotubes with diameters of less than 2 nm serve as pores is being used to develop desalination membranes. The fundamental advantage of this technology is based on the large enhancements of the fluid and gas flow in carbon nanotube pores. Carbon nanotubes can be viewed as seamless, atomically smooth carbon "straws" whose diameters range from less than 1 nm to tens of nanometers (a water molecule is ~0.3 nm). Water flows through these unique pores 1,000 times faster than through any other pore of similar diameter. Moreover, gases also flow through the nanotubes' pores more than 100 times faster than through any other nanometer-scale pore. This reduction in flow resistance manifests itself in large enhancements of the membrane permeability and in drastic reduction of viscous losses. The membranes, which are being developed by Porifera, promise superior performance (permeability, selectivity, and durability) at costs similar to those of membranes currently used in liquid phase separations and require less energy to operate. The system also allows for three important membrane characteristics — pore size, pore entrance and exit characteristics, and membrane matrix — to be modified independently, meaning that the membranes can be tailored for improved performance in specific applications. (Source: *Porifera: Carbon Nanotube Membranes*, available at http://www.poriferanano.com.)

- *Improved efficiency of existing membranes.* IBM is developing a nano membrane made from fluorine materials that bond with water on contact and allow it to flow freely, but when salt and other impurities come in contact with the same membrane, a highly acidic reaction takes place that results in impurity ionization and hence repulsion or separation. (Source: *Green Technology — IBM's Silicon Chip Material Science to Treat Water*, available at http://green.tmcnet.com/topics/green/articles/52594-ibms-silicon-chip-material-science-treat-water.htm.)

- *Nano-based filters to inhibit corrosion in recirculating systems and protect RO membranes from fouling.* Argonide Corporation (no relationship to Argonne National Laboratory) has developed a nanofiber filter that can filter particles as large as tens of microns or a small as a few nanometers (submicron and colloidal). These NanoCeram® filters, which are based on the principle of electrostatic attraction and adsorption, can be used for a variety of purposes, including industrial water purification and prefiltration for RO membranes. They remove small particles, colloidal iron, silica, chlorine, bacteria, and viruses, and can be adapted to remove specified other contaminants from water.

The filters use nanoaluminum oxide fibers that are 2 nm in diameter and 200–300 nm in length, which are attached to microglass strands. With a surface area of about 500 square meters per gram, the fibers produce an electropositive charge when water flows through them. Most impurities carry a slight negative charge and are thus absorbed by the nanoalumina. A single layer of the nonwoven filter media, although it has a relatively large pore size of about 2 microns, is capable of removing more than 99.99 percent of 0.025 micron particles. (Three layers remove up to 99.9999 percent) at flow rates more than 200 times faster than virus-rated membranes that remove particles by traditional mechanical filtration. Because the filter's pore size is much larger compared to other membranes, it is also less susceptible to clogging by small particles. Competitive nonwoven filters of the same pore size can match the flow rate of this filter, but they are less efficient, particularly for sub-micron particles. A recent modification to the original design incorporates a powdered activated carbon, which results in high-performance removal of chlorine. The dynamic adsorption capacity of this material is more than 100 times that of carbon-containing non-woven media that use granular activated carbon. These filters can remove a range of soluble and insoluble contaminants, whereas most water filter systems have separate sediment and activated carbon stages and often require separate housings. An even newer filter embodies a silver-containing agent that substantially reduces the growth of bacteria that would minimize if not eliminate the need for chemical bactericides, many of which can initiate corrosion. Argonide notes that the nanoalumina media can retain virtually any nanopowder desired, meaning that it can be adapted to remove specific contaminants from water, depending on the specific needs of the application.

The NanoCeram® filtration technology is being used commercially for several applications, including reducing the size of Toyota's "water footprint" in the United States. Toyota wanted to reduce its water consumption from 900 gallons per car to 300 gallons per car. To purify recycled water, Toyota uses RO membranes, which foul easily. Even with the use of standard prefilters, the RO membranes at one plant needed to be replaced every 2–3 months. By substituting the NanoCeram® filters, the lifetime of the filters was extended to one year. Toyota also used the NanoCeram® filters to solve a longstanding corrosion problem, which was leading to the premature decay of its extensive chilled water pipe network. By using the filters to cleanse the roughly 16 miles of chill pipes in a typical plant, the cause of the corrosion problem — iron oxide and the bacteria that feed on it — was removed.

Argonide notes that its filters would facilitate the use of nontraditional water sources in power plants. For example, waters from acid mine drainage and mine pools tend to be acidic and laden with metal contaminants. The Argonide filters remove particulates at pH levels down to about 3.5, and lime can be added to raise the pH to 3.5. The filters will also remove limited amounts of soluble metals, which may be adequate where the soluble metals are in the low parts per million (ppm) or ppb range. When using treated municipal water, the filters could be used as a polisher before RO treatment.

Argonide is poised to conduct a pilot test at a coal-fired power plant to demonstrate utility of its filters for one or both of the following applications: (1) purifying water in a recirculating system to inhibit corrosion; and (2) protecting the RO system that may be used when using brackish or seawater for make-up feedwater. It has licensed the rights to manufacture, market, and sell its filter media to Ahlstrom Filtration, which mass produces the media at a plant in Pennsylvania. Consequently, there would be no scale-up issues regarding manufacture of the filter media or by other raw materials. Argonide is interested in working with and obtaining funding from DOE to conduct this pilot demonstration and estimates that the cost of such a pilot would be less than $100,000. The proposed pilot study is summarized below.

Proposed Argonide Pilot Demonstration. The approach would be to install a bypass loop in the recirculating system and place coupons of the structure metal piping on the cold side of the loop. Argonide proposes a 240-gallons-per-minute (gpm) water purification system consisting of one each of a prefilter (1 micron) and the NanoCeram® (noncarbon) filter that would have an added bacteriastatic agent in the second 240-gpm unit. Controls would be run without filtering, and the test coupons would be examined for metal corrosion. The NanoCeram® system would then be run without prefiltering. Its ability to collect dissolved metals and other matter could be sufficient. Metallographic evaluation of the coupons plus assay of the colloidal matter and for bacterial contamination would provide the desired information. Changes in the turbidity of the fluid both upstream and downstream of the filters would also be measured. If filter life is too short, the 1-micron prefilter would then be employed. The program would be conducted over a one-year period during which time a replacement of filters could be required. Argonide would be the primary contractor for installation and for performing the study.

5.1.2 Membranes for CO_2 Capture

As noted in Chapter 4, most post-combustion CO_2 capture technologies involve absorption, which requires significant amounts of energy and water. Research is well underway on the use of nanomembranes to capture CO_2. Similar to the membranes used for water filtration, membranes used to separate gases are physical barriers that allow some gases to move through much faster than other gases. Membranes for CO_2 capture, which have small holes that approach the size of the molecules, let smaller molecules through more rapidly. With proper nanoengineering of the molecular structure to favor selective adsorption and surface transport along the pore walls, membranes that favor larger molecules over smaller ones can also be produced. A European project, "Nanomembranes against Global Warming" (NanoGLOWA)[2] reports that by using nanomembranes rather than absorption technology, the energy requirements for CO_2 capture could be reduced from about 25 percent of the energy produced at the power plant to about 8 percent. NanoGLOWA researchers are simultaneously designing three main types of membranes: (1) polymer membranes (diffusion transport membranes, fixed-site carrier membranes, and ionomeric high voltage membranes); (2) carbon molecular sieve membranes; and (3) ceramic membranes. Polymeric membranes are relatively inexpensive, but selectivity (and hence efficacy) may be reduced at higher pressures. The membranes are scheduled to be pilot-tested in four large European power plants in 2011. More information on the nanomembranes being developed by NanoGLOWA is available at http://www.nanoglowa.com/cm.html.

---

[2] NanoGLOWA consists of 25 participants from universities, power plant operators, and industry in 12 European countries, brought together to develop optimal nanostructured membranes and installations for CO_2 capture from power plants. The five-year project, which began in 2006 and is scheduled for completion in 2011, is funded by the European Commission.

Researchers at the Norwegian University of Science and Technology have patented a new type of membrane made from a plastic material that has been structured by means of nanotechnology to catch CO_2 while letting other waste gases pass freely. Its effectiveness increases proportionally to the concentration of CO_2 in the gas. The method is the facilitated transport method, and it is based on the way the human lung removes CO_2 — a complex but effective mechanism. Rather than using a filter that separates CO_2 directly from other molecules, this approach uses an "agent," i.e., a fixed carrier in the membrane that helps to convert and transport the gas to be removed. A reversible reaction occurs at the amine fixed site carriers forming bicarbonates (HCO_3) from CO_2 and water molecules. The bicarbonates then move to the other side of the membrane and release CO_2 while the other gases are retained by the membrane. See http://www.engineerlive.com/Oil-and-Gas-Engineer/Environment_Solution/New_membranes_design_will_improve_carbon_dioxide_capture/19942/ for more information on this membrane.

Researchers at the University of Connecticut have received a grant from the National Science Foundation to begin building a prototype device to test a process that would attach a specific enzyme to nanoparticles, which would trap the CO_2 molecule, after it is created but before it is released to the atmosphere. The enzyme-laden nanoparticles would then convert the CO_2 from the flue gases of the power plant into water-soluble bicarbonate (HCO_3). (Source: http://advance.uconn.edu/2008/080219/08021909.htm.)

General Electric is currently working on minimizing performance degradation over long periods of time as well as making membrane units large enough to handle CO_2 at the scales produced by a power plant.

5.2 COATINGS AND LUBRICANTS

Nanotechnology-based coatings have been and are being developed to help insulate and reduce heat loss to increase overall operating efficiency; to limit and prevent corrosion in utility boilers; to improve corrosion, erosion and wear resistance in boiler tubes and dry FGD scrubbers; and to improve fuel efficiency. Also, effective lubricants that can withstand severe chemical and physical conditions have been developed with nanotechnology. Examples of these coatings and lubricants are described in the following paragraphs.

5.2.1 Coatings to Insulate and Reduce Heat Loss Throughout the Plant

Nanotechnology-based coatings can be used to insulate a variety of equipment in a coal-fired power plant to increase operating efficiency and reduce energy (and water) use. Industrial Nanotech, Inc., has developed and is marketing a nanotechnology-based coating called Nansulate® that is designed to increase operating efficiency by reducing heat loss and equipment maintenance requirements. The Nansulate coatings incorporate a nanocomposite material called Hydro-NM-Oxide, which has very low measured thermal conductivity. The internal nanoscale architecture of the material consists of a network of tunnels, which help to reduce the heat energy that is flowing through it. The unique makeup of the nanocomposite allows the material to insulate in a thin layer. The material insulates by means of low thermal conduction (or by reducing conduction of heat energy through the material).

In a typical power plant, heat (and hence energy) loss can occur through the exterior of the boiler, steam pipes, the turbine chamber, and heat exchangers. By painting an insulation coating onto each part of the heat generation flow, the entire system becomes more energy-efficient, and less energy is needed to reach the same process temperature. The coating offers protection for large equipment and surfaces that are not easy to insulate with traditional means (e.g., fiberglass). Traditional forms of insulation are also subject to moisture infiltration, which reduces their insulating benefit and contributes to corrosion of the underlying surface (corrosion under insulation), necessitating regular inspection and replacement of pipes and other equipment. The Nansulate coating, which combines the characteristics of extremely low thermal conduction and hydrophobicity into a water-based acrylic latex, provides thermal insulation, corrosion resistance, and mold resistance in a thin clear layer (which allows for visual inspection of the surfaces without removal). The coating can be used in applications up to 400 degrees Fahrenheit (F) and can be used to insulate equipment that currently may not have a suitable insulation option such as boilers (so that less fuel is used to produce the same amount of steam energy), feed water heaters (to reduce the amount of steam needed to heat the water to the desired temperature, and also protect the exterior from corrosion); furnaces; and other equipment, which because of size or area limitations, cannot be insulated with conventional insulting materials. The coating can also be used to replace traditional insulation and can be used in (1) the pre-treatment area/furnace where coal is heated to reduce moisture content, causing less energy to be used to maintain desired pre-treatment temperature; (2) the chamber that holds the steam turbine generator to reduce heat loss through the chamber walls; and (3) steam pipes, heat exchangers and similar system components, so that more of the energy produced by the boiler is delivered to the end power-generation process and less is lost through the walls of the steam pipes and other equipment.

5.2.2 Coatings to Limit/Prevent Corrosion in Utility Boilers

In 2007, EPRI issued a report, *State of Knowledge Review of Nanostructured Coatings for Boiler Tube Applications,* which provided a detailed review of nanostructured coatings and assessed the potential for protection of boiler tubes by nanostructured coatings (EPRI 2007). The study reported that on the basis of laboratory experiments, nanostructured coatings are harder, tougher, and more resistant to high-temperature oxidation and corrosion than their counterpart conventional coatings, and that nanostructured coatings containing chromium and/or aluminum are much more resistant to oxidation and corrosion than conventional coatings containing these substances. Slower-reaction kinetics with high-temperature environments stem from the establishment of thin, impervious, thermally grown oxides on the nanostructured coating surfaces. Short-circuit diffusion of aluminum (Al) or chromium (Cr) through the grain boundaries is considered the mechanism for promoting selective oxidation by nanocrystalline grains. For this reason, nanostructured coatings require only about one-fourth the aluminum or chromium content needed in conventional coatings to establish thin, continuous, protective thermally grown oxides. Further, the thermally grown oxides on nanostructured coatings are more adherent and more resistant to spalling (where fragments of material are ejected from a body due to impact or stress) than are oxide scales on conventional coatings. Nanostructured coatings are also more erosion-resistant and better bonded to substrates than conventional coatings (EPRI 2007).

In 2008, EPRI began an effort to improve the reliability and availability of fossil-fired ultra supercritical boilers and oxy-fuel combustor systems by developing advanced nanostructured coatings that are optimized by using computational modeling and are validated via experimental verification and testing in simulated boiler environments for three different coal conditions and temperatures (EPRI 2008). The idea is that power producers will be able to use the results of this evaluation of advanced "MCrAl" nanocoating systems to determine which coatings to consider for extending boiler tube service life where corrosive conditions are a concern. ('M' refers to iron, nickel, and cobalt)

Research has found that the performance of MCrAl coatings can be enhanced significantly when coatings are deposited by using advanced processing techniques that produce nanoscale microstructures. For a given Cr or Al concentration, nanoscale coatings exhibit longer life than the conventional coatings. However, the chemical composition and processing parameters need to be optimized for these advanced coatings — either by trial-and-error methods (which can be costly and time-consuming) or by using computational methods. The computational modeling effort has shown, among other things, that nanocoatings offer improved properties over conventional coatings with coarse grains and that MCrAl type coatings can be applied in nano form as iron-aluminum (Fe-Al) and nickel-aluminum (Ni-Al) coatings.

5.2.3 Coatings to Improve Corrosion, Erosion, and Wear Resistance

The NanoSteel Company has developed several nanotechnology-based alloys that combine high hardness/toughness properties with greatly enhanced corrosion, erosion, and wear resistance. When applied as thermal spray coatings, the alloys form a microstructure with grain sizes refined to a nanoscale. Example applications include (1) boiler tube coatings to resist erosion and corrosion on waterwalls and super heater, reheater, and economizer tube assemblies; and (2) applications for dry scrubbers. With dry scrubbers (a technology for removing SO_2 from flue gas), calcium sulfite, calcium sulfate, and chloride deposits can build up on the walls of the chamber wall where fly ash, lime milk and cooling water are sprayed to absorb SO_2 from flue gas. This process causes corrosion and pitting damage on the carbon steel chamber wall, which can cause the chamber to buckle and fail. The thermal spray coating, when applied to the area where the process spraying occurs, creates a metallic barrier between the chamber wall and the corrosive build up, providing resistance to corrosion and pitting damage. The coatings can also be applied to other dry scrubber components such as turning vanes, spray nozzles, and piping to resist erosion. (Source: www.nanosteelco.com.)

5.2.4 Coatings to Improve Reliability and Fuel Efficiency

As mentioned in Chapter 4, undesirable corrosive and fouling side effects of burning fossil fuels and the side effects of operating NO_x pollution control devices can create problems within regenerative air heaters. A recently developed coating, DuraTEC™ tempered enamel coating, is a nanotechnology-tempered vitreous enamel coating that incorporates metal oxide nanoparticles into a ceramic matrix that is applied to the element sheets of the power plant's regenerative air heaters. The coating provides a nonstick surface that eliminates ammonium bisulfate deposition and increases the life expectancy and durability of the air heater over that of bare steel or traditional standard enamel. The coating can also be embedded with an SCR type of catalyst to reduce the quantity of NO_x emissions from the power plant below the level achieved with the existing SCR. The coating uses various application techniques (wet enameling, dry electrostatic enameling, and a combination of both) in conjunction with nanotechnology to produce a strong amorphous bond with the steel. The enamel and steel "melt" together without delineation between the two layers. The advantage of this amorphous bond is that if the top layer of enameling is lost, the underlying steel is still protected and resistant to corrosion. For more information on this coating, see http://www.paragonairheater.com/duratec.htm.

5.2.5 Lubricants with High Chemical and Physical Stability

Under severe conditions, traditional lubricants can be squeezed from contact areas. The addition of nanoparticles to the lubricants can reduce friction and wear rates and increase load-bearing capacity. Nanotechnology-based extreme-pressure and anti-wear additives have been found to have high chemical and physical stability under extreme conditions, meaning longer equipment operation, increased efficiency, and extended maintenance intervals (Kennedy 2010). Because the nanoscale particles infiltrate the tiniest spaces between contacting surfaces, coverage is improved. The nanoparticles can also be impregnated into polymer or metal coatings to provide anti-friction properties, and into porous metal parts to make self-lubricating components. ApNano Materials makes a nanotechnology-based lubricant called NanoLub™, in which nanoparticles of tungsten disulfide (WS_2) that are structured as nested spheres called inorganic fullerenes, are blended into traditional motor oil. (A fullerene is a nanotube in the form of a hollow sphere.) The lubrication method involves the slipping off of the layers under loads to form an adherent film that reduces friction and wear. The nanoparticles enhance the lubricating properties of the oil or grease with respect to wear and friction by an order of magnitude over the same lubricant without the nanoparticles. (Source: http://www.apnano.com/.) The company is also developing a nanotechnology-based approach for obtaining self-lubricating hard (dry) coatings such as cobalt, by impregnating the WS_2 fullerenes in the coatings. (Many machine parts are made of cobalt coatings.) The coating acts as a reservoir for nanoparticles, which, when slowly released from the surface, provide easy shear and reduced oxidation of the coating or native metal surface.

5.2.6 Welding Materials

Nanotechnology has been used to produce unique, ultrafine microstructures that result in properties that are superior to those achievable by standard welding metallurgy. A proprietary nanotechnology is being used in a product that has recently become commercially developed to provide abrasion and erosion resistance performance that is similar to tungsten carbide but uses a lower cost ferrous alloy with bulk hardness values over 70 HRC. (HRC is a measure of hardness; very hard steel, such as that used in a good knife blade, has an HRC 55–62.) (Source: www.nanosteelco.com.)

5.3 CATALYSTS AND ENZYMES

Because of the reactive properties of nanomaterials, which are due in large part to their relatively large surface area when compared with their bulk counterparts, their use in catalysts is a significant RD&D area. Two types of catalysts with potential applicability for reducing water use via improved plant efficiency (and reduced energy and water consumption) are described here. These are nano-based enzymes to improve fuel efficiency and catalysts to improve the efficiency of sulfur removal in wet FGD scrubbers.

5.3.1 Enzymes to Improve Fuel Efficiency

Enzymes are biomolecules that catalyze, or increase the rates of, chemical reactions. Enzymatic reactions can be millions of times faster than those of comparable reactions without enzymes. A new generation of enzymes has been developed, which, when applied to coal at ambient temperatures and pressures prior to combustion, will increase the rate of reaction and thus, the efficiency of combustion. These enzymes also react with the nitrogen and sulfur in the coal to reduce emissions of SO_2 and NO_x. Earlier generations of the enzymes were too large to penetrate the coal sufficiently to result in a significant change in combustion. However, the new enzymes, in which the particles are sized at the nanoscale level, are small enough to penetrate the pore space of the coal and hence, have the desired effects of increased combustion efficiency and reduced pollution generation. One company, Simpert Technology, has found that the application of the nanoenzyme in a test facility resulted in a 5 to 15 percent reduction in coal consumption, a 10 to 29 percent reduction in combustion emissions, reduced flyash generation, less scale on boiler tubes, and (because of lowered coal consumption), reductions in the physical quantities of pollutants such as arsenic, cadmium, chlorine, mercury, nickel and phosphorous. In late 2009, the company completed a commercial scale-up project at a combined heat and power plant in the Midwest that burns low-sulfur bituminous coal. The nanoenzyme was applied as the coal was unloaded. (One liter when diluted with 3,000 liters of water can treat 1,000 tons of coal.) Combustion progress was monitored in real time, and data indicated that the nano enzyme biocatalysts and plant operation optimization can achieve an eight percent improvement in energy efficiency. The data also indicated a 35 percent reduction in NO_x and a 16 percent reduction in SO_2. For more information on the nanoenzymes for coal, see http://www.simpert.com/.

5.3.2 Catalysts to Improve Efficiency of Sulfur Removal

Wet scrubbers are considered state-of-the-art methods for achieving SO_2 removal efficiencies in the range of 90 to 98 percent. Dry scrubbing with lime slurries results in lower efficiencies and is generally considered useful for lower-sulfur coals. However, when combined with sorbent injections, such as metallic oxides and metallic salts, dry scrubbing may become more viable for higher-sulfur coal applications. Recently, research has been conducted on the use of nano titanium dioxide, nano-TiO_2, as a catalyst to improve sulfur removal efficiency with dry scrubbers (Zhao et al. 2009). Zhao and his colleagues conducted some fundamental experiments to study the effects of nano-TiO_2 additives on the lime's (CaO) desulfurization efficiency, coal combustion efficiency, and the mechanisms of the additive's effect. In their experiments, Zhao and his colleagues found that the conversion of lime increased from 36 percent to 70 percent when the dosage rate of nano-TiO_2 increased from zero percent to eight percent, suggesting that desulfurization efficiency increased with the increased conversion of lime, and that was a catalyst in the combustion. The research also showed that the maximum efficiency occurred when the ratio of calcium to sulfur was 2 to 1, the temperature was 850 degrees Celsius (C), and the dosage of nano-TiO_2 was six percent. With these conditions, desulfurization efficiency was 87.7 percent, or 16.8 percent higher than without the TiO_2 addition. Zhao et al. concluded that these findings are helpful in understanding and developing this new combustion additive and that the research may provide a useful basis for further application of the additive in coal-fired power plants. See Zhao et al. (2009) for more information on this research.

5.4 MATERIALS THAT CAN WITHSTAND HIGH TEMPERATURES

As noted in Chapter 4, higher temperature operations in coal-fired power plants have been limited because of the inability of many materials to withstand high temperatures, including wear and chemical degradation. One category of material that has been investigated for use under conditions of high temperature, high wear, and chemical attack are ceramics. Because their inherent brittleness has limited their use, much research has been directed toward making ceramics more tolerant. This research has included designs that incorporate fibers and particles that deflect cracks. A more recent and promising approach involves the use of ceramic nanocomposites. Nanocomposite materials encompass a variety of systems including one-dimensional, two-dimensional, three-dimensional and amorphous materials, made of distinctly dissimilar components and mixed at the nanometer scale (Tomar 2010). Nanocomposites are being researched because they offer several improvements in the properties over the simpler structures. These include, but are not limited to (1) improved mechanical properties such as strength and dimensional stability; (2) decreased permeability to gases, water, and hydrocarbons; (3) higher thermal stability and heat distortion; and (4) higher chemical resistance. The Multiphysics Lab at Purdue University is conducting research on the performance of one class of nanocomposites — silicon carbide/silicon nitride composites, which after being embedded with particles in the 20–300 nm range, have demonstrated significant improvements (200 percent) in strength and fracture toughness at high temperatures. This research is aimed at understanding the performance of these nanocomposites in the extreme environments found in power generation cycles and in understanding thermal conduction and thermal issues in materials for thermoelectric power generation (Tomar 2010).

5.5 NANOFLUIDS

A noted in Chapter 4, conventional heat transfer fluids (e.g., air, water, oil) have inherently low thermal conductivities compared with solids. The information in this paragraph, which describes early research on the potential use of nanofluids — heat transfer fluids engineered by dispersing and stably suspending nanoparticles with typical lengths of about 10 nm in traditional heat transfer fluids — for cooling, comes from Choi (2008). For more than a century, scientists have tried to enhance the poor thermal conductivity of liquids by adding solid particles. However, until very recently, these studies had been limited to using millimeter- or micrometer-sized particles, which tend to settle rapidly in liquids; require high concentrations to obtain appreciable improvements in the thermal conductivity; and cause abrasion, clogging, and pressure drop. In the 1990s, Choi began exploring ways to apply nanotechnology to heat transfer engineering, and in 1995, he introduced the possibility that with nanofluids, convection heat transfer coefficients could be doubled, which would increase pumping power by a factor of 10 — significantly increasing efficiency. Since then, experiments have shown that nanofluids exhibit novel thermal transport phenomena. For example, a very low concentration of copper nanoparticles (0.3 percent by volume) suspended in ethylene glycol showed a 40 percent increase in the thermal conductivity of the fluid; the thermal conductivity of nanofluids is strongly temperature-dependent; and critical heat flux can be significantly increased by increasing fluid pH.

These and other discoveries regarding the enhanced thermal conductivity of nanofluids have significant implications for more efficient cooling systems, resulting in higher productivity and energy and water savings. Most of the research to date has focused on determining the levels of thermal enhancement of a variety of nanofluids and investigations of the effects of various parameters, such as particle material, shape, size, and volume concentration; base fluid; temperature; and pH. However, some recent research has been aimed at potential applications for use in commercial and industrial settings. The information in the remainder of this paragraph comes from Singh et al. (2009). Singh and his colleagues recently conducted a study to investigate a commercially available silicon-carbide (SiC)/water fluid to assess the viability of the fluid for an industrial application. SiC was used because it is relatively easy to incorporate into a fluid, has good long-term stability, has a high thermal conductivity among ceramics, and is readily available. The researchers conducted physical and microstructural characterizations of the nanofluid for particle size, viscosity, and agglomeration, and they investigated the effects of particle loadings and temperature on viscosity and thermal conductivity. The results were used to correlate the physical properties to the thermal behavior to assess the viability of the fluid for an industrial application. The results showed that the SiC nanofluid system exhibited increased thermal conductivity and viscosity as a function of particle loading at ambient room temperature.

A subsequent study reported the first experimental heat transfer results for the SiC/water nanofluid (Yu et al. 2009). The information in the remainder of this paragraph comes from Yu et al. (2009). The experiment compared the results of a 3.7 percent volume-percent SiC water nanofluid with those for the base fluid water with those from heat transfer predictions. The experiment also compared nanofluids on the basis of the combined effects of heat transfer enhancement and pumping power increase. The experiment found that the heat transfer enhancement of the nanofluid was 50 to 60 percent higher than water. The researchers explain that the concept of pumping power penalty is often used to compare augmented heat transfer situations, and they combined pumping power with heat transfer enhancement to produce a parameter (ratio of heat transfer enhancement to pumping power increase) that would indicate the overall merit of a nanofluid. The ratio for the SiC nanofluid was more favorable than that for an alumina (Al_2O_3)/water nanofluid (0.8 compared with 0.6), meaning that the SiC fluid demonstrated a greater gain in the heat transfer enhancement compared to the pumping power penalty than the alumina nanofluid. (Al_2O_3 is a typical commercially available polycrystalline ceramic with relatively good thermal conductivity.) The results showed that the "SiC/water nanofluid is well behaved, the thermal conductivity enhancement is reasonably high, and the viscosity increased is relatively low.[3] Settling and agglomeration do not occur, and all of these conditions contribute to the potential commercial viability of the fluid. This result points to the direction of continued research in the development of a fully viable water-based nanofluid for commercial applications" (Yu et al. 2009).

In summary, water-based nanofluids have enhanced thermal conductivity, but they also have increased viscosity, meaning that because more power is required to pump them through a system, there is a tradeoff between increased thermal conductivity (that could lead to increased heat transfer) and increased energy for pumping. According to Routbort (2010), research efforts underway to address this issue include using heat transfer fluids that have a higher viscosity than water (e.g., ethylene glycol and water), adjusting the pH, changing particle size and shape, and coating the particles. In recent experiments, for example, researchers were able to lower the viscosity and in a SiC/ethylene-water system achieved a 14 percent increase in heat transfer coefficient using larger SiC particles. At the same time, if more heat could be removed, the pump speed could be lowered to remove the same amount of heat as the base fluid, thereby saving energy and not altering the viscosity. On the basis of the above research, there is reason to believe that given more attention, nanofluids could be used to increase efficiency and reduce water consumption in coal-fired power plants.

5.6 NANOSENSORS

Nanosensors, or sensors made of nanomaterials, can be extremely sensitive, selective, and responsive. As such, they could be smaller and cheaper, and consume less power than conventional sensors. Sensors and controls that are small in size; work safely in the presence of electromagnetic fields, high temperatures, and high pressures; and can be changed cost-effectively may provide the ability to monitor conditions in the infrastructure and monitor for pollutants continually.

The unique properties of nanostructures such as nanoparticles, carbon nanotubes, nanowires, nanoscale thin films, semiconductor quantum dots, and nanocantilevers lend themselves to sensing applications. In particular, their small size, light weight, and large reactive surface areas have been shown to improve — by orders of magnitude — the sensitivity, selectivity, and response time of sensor technologies, and to dramatically reduce the size, weight, and power requirements of the resulting monitoring devices when compared to conventional, macroscale alternatives (Shelley 2008). Nanostructured materials can enhance sensor applications that depend on sorption of the target analyte (the specific chemical or biological species of interest), because sorption depends on surface areas and surface chemistry, both of which are enhanced at the nanoscale. Nanotechnology offers the ability to increase selectivity of monitored constituents; by modifying or functionalizing the surfaces of nanotubes, (e.g., by adding coatings), they can be made more selective to specific analytes. Sensor interfaces made from nanoporous or nanocrystalline materials increase the surface areas significantly, thereby increasing the signal intensity from the sensor.

---

[3] Viscosity is important, because increased viscosity can mean increased energy requirements for pumping, which would need to be weighed against the increased cooling efficiencies.

Nanotechnology also offers the ability to detect multiple analytes simultaneously, by using tens or hundreds or thousands of minute sensors within a single monitoring device. Individual sensors are tuned for a specific agent, by using a combination of nanowires or nanotubes that contain different chemical coatings or functional groups. Powerful signal processing and pattern recognition algorithms allow for the generation of a unique fingerprint that identifies target molecules (Shelley 2008). The use of conventional sensors to monitor numerous target gases is a much more bulky and costly alternative. It is anticipated that the availability of ultra-compact, low-power nanosensors-based monitoring devices will result in large sensor arrays that can simultaneously monitor a large number of analytes in a single application with increased reliability, sensitivity, and accuracy. Future sensor systems can be envisioned that will include some combination of sample collection, transducers able to sense multiple analytes, sophisticated analysis electronics, and wireless communications systems, and nanotechnology offers new solutions for the technical challenges associated with each of these. In addition, nanofabrication processes are enabling the development of high-density arrays of transducers, and nanoelectronics will process and analyze the signals coming from these arrays.

Sensors in uncontrolled locations become contaminated by a variety of substances including volatile organic vapors, carbon soot, and oil vapors, as well as dust and pollen. A self-cleaning function capable of oxidizing contaminants would extend sensor lifetime and minimize sensor errors. Researchers have recently developed sensors made from titanium nanotubes coated with a discontinuous layer of palladium. The photocatalytic properties of titanium nanotubes are so large — a factor of 100 times greater than any other form of titanium — that sensor contaminants are efficiently removed with exposure to ultraviolet light, so that the sensors effectively recover or retain their original sensitivity to hydrogen (Science Daily 2004).

There are at least five classes of nanotechnology-based transducers — the material or device that converts a recognition event into a measureable signal. These are electro/chemical, electromagnetic, spectroscopic, magnetic, and mechanical. For more information on the status and future directions various nanotechnology-based sensing systems and nanotechnology enabled transducers, see Evans (2009). Numerous nano-enabled sensor designs are being developed to improve detection and monitoring of corrosion, chemicals, leaks, industrial processes, environmental monitoring, and other applications. Presented below are four examples of recent research into nano-enabled sensors that could help support water reduction at coal-fired power plants.

5.6.1 Strain and Impact Damage Identification

Research has been conducted on carbon nanotube-based "sensing skins" to detect strain and damage (e.g., corrosion) in underlying structural materials. The following information comes from Loh et al. (2009). SWNTs, whose characteristics include strength, large surface area, high aspect ratio, and mechanical durability, are well suited for chemical functionalization. That is, by binding molecules to the surface of SWNTs, electromechanical and electrochemical sensing transduction mechanisms can be encoded into a carbon nanotube composite. In this research, Loh and his colleagues employed SWNTs as building blocks for the design and fabrication of multifunctional sensing skins that are capable of monitoring structural damage without having to probe multiple discrete sensor locations to infer the characteristics (type, location, and severity) of damage. The skin is intended to provide a spatial image of the deformations and identify and locate breaks in the skin that could be attributed to cracking of the underlying structure or to physical impacts on the skin. When combined with a spatial conductivity mapping technique, the skins were validated for distributed strain and damage sensing. The skins can be coated onto structural surfaces to monitor distributed damage processes. The authors report that they plan to validate the skins for damage detection on actual structural components and to test long-term sensor stability tests in which the skins will be exposed to ambient conditions to identify potential environmental factors that could affect sensor performance.

5.6.2 Water Quality Detection and Monitoring

Cost-effective detection and monitoring of constituents before and after treatment will help improve the efficiency of water treatment technologies in coal-fired power plants. Researchers have recently demonstrated that simple, high-performance biosensors can be made from common filtration papers impregnated with SWNTs to detect toxins in water (Wang et al. 2009). The result is a simple, rapid, versatile, and inexpensive method for detecting toxins. The approach is based on the electrical conductivity of the SWNTs and the distance between the nanotubes, with the conductivity of the network exhibiting a strong dependence on the presence of the analyte. In their research, the scientists dispersed an antibody to a particular toxin (microcystin-LR, or MC-LR) together with SWNTs, and then dip-coated a porous fibrous material (paper) with the dispersion to render it conductive. The change in conductivity of the paper was used to sense the MC-LR in the water rapidly and accurately. The detection limit was comparable to traditional methods of MC-LR detection, but it reduced the time of analysis by more than an order of magnitude, thus making it much more amenable to practical applications. The authors conclude that similar preparations can be used for various other rapid sensors, including detecting pollutants in waters or gases at coal-fired power plants (Kotov 2010).

5.6.3 Measuring Mercury in Flue Gas

The ability to optimize the energy (and water) expended to control mercury emissions can be enhanced by sensors that can accurately detect and measure the pollutant. Researchers at RMIT University in Australia (Sawant et al. 2009) have developed nano-based mercury sensors that can accurately measure the amount of mercury in effluent gas streams. The accuracy of existing mercury sensors can be affected by the volatile organic compounds (VOCs), ammonia, and water vapor that exist in industrial chimneys. The new sensor is sensitive enough to provide precise readings of the amount of mercury vapor in the emissions and is also robust enough to cope with the harsh environments. The nano sensor was developed by using a nano-engineered gold surface that was altered by using a patented electrochemical process to produce hundreds of nano spikes, which were then used with a finely tuned set of scales that measure weight at the molecular level to determine the levels of mercury in the atmosphere. While it is well-known that gold and mercury attract each other, a regular gold surface does not absorb much mercury vapor, and measurements are inconsistent. The nano-engineered gold surfaces are 180 percent more sensitive than the non-modified surfaces. The researchers note that the surfaces have worked successfully at a range of extreme temperatures over many months.

5.7 BATTERIES AND CAPACITORS

Nanotechnology is being used in the development of large-scale batteries and capacitors to increase available power and decrease recharge time. For example, by coating the surface of an electrode with nanoparticles, the surface area of the electrode is increased, which allows more current to flow between the electrode and the chemicals inside the battery. By improving the design of finely layered compounds for storing high densities of lithium ions and minimizing the distance the ions must travel, nanotechnology facilitates faster charging and discharging. Because the internal structure of a battery and a capacitor can be similar, researchers are investigating methods for using nanostructured oxide materials to combine the best characteristics of both in devices that can deliver power fast (as with a capacitor) but also sustained over time (as with a battery) (Bell 2009). As noted in Chapter 4, the ability to store energy produced during low demand periods and release it during periods of higher demand allows a power plant to operate at a constant, efficient load. At least three companies are developing technologies that could lead to large-scale utility storage that uses nanotechnology. Highlights of these technologies are summarized below.

- *Improved battery performance.* Altairnano has replaced traditional graphite materials used in conventional lithium-ion batteries with a proprietary, nano-structured lithium titanate. The formulation of the material provides for a substantially more spherical material, resulting in improved battery performance. Altairnano recently received a $100,000 grant to conduct research on the surface modification of electrode coatings in battery cells with the objective of increasing temperature and cycle life performance by decreasing reaction rates with the electrolyte. The company's lithium titanate batteries are scalable so that applications include complete energy storage systems for use in providing frequency regulation (regulation of the stability of the electric grid) and renewable integration for the electric grid. Other performance attributes include fast charge and discharge rates, long cycle life, long calendar life and the ability to operate under diverse environmental and temperature conditions. (Source: www.altairnano.com.)

- *Increased power density.* A123 Systems has a nanotechnology-based lithium-ion battery that uses low-impedance nanophosphate electrode materials. Thousands of times smaller than the micron-sized materials used in first-generation lithium-ion batteries, the nanophosphates allows for twice the power density of competitive products. The cells and electrodes are designed to deliver low cost per watt and cost per watt-hour performance. They have a higher voltage than other long-life systems, which leads to reduced lifecycle and system costs resulting in greater overall price performance. A123 Systems has an energy storage system called the Hybrid Ancillary Power Unit, which is designed to provide backup services by storing energy until it is needed by the grid in the event of a power plant or other asset failure. This large battery (the size of a tractor trailer) "hybridizes" a power plant by adding a multi-megawatt energy storage system to the plant. It increases the capacity, responsiveness, and efficiency of an individual power plant and provides high-quality power within milliseconds of a demand signal. It scales from 2 to 200 megawatts of power. The system frees up plant capacity reserves that are typically set aside for ancillary services to enable the utility to focus on baseline electricity generation. A123 systems reports that independent third-party studies have shown that by using high-efficiency energy storage systems such as the Hybrid Ancillary Power Unit for frequency regulation, associated emissions of CO_2, SO_2, and NO_x can be reduced by as much as 80 percent over traditional power plant ancillary services. (Source: www.a123systems.com.)

- *Improved fabrication to increase functionality.* Planar Energy has developed a battery fabrication method that uses nanotechnology in a deposition process that allows solid-state electrolyte materials to be stacked as thin films directly on active layers in the battery. This approach eliminates the historic process of having to deposit films on separate substrates and then mechanically joining them. By allowing for the direct growth of self-assembled films directly on flexible substrates or on top of other films, the process simplifies the battery manufacturing process and enables existing battery chemistries to function at 95 percent of their theoretical value. The batteries do not discharge, allowing them to sit for long periods of time while retaining their charge. (Traditional lithium-ion batteries have high discharge rates.) (Source: http://www.planarenergy.com/technology.html.)

6 CONCLUSIONS AND RECOMMENDATIONS

A variety of potential nanotechnology-based applications have been identified that could help — either directly or indirectly — reduce freshwater consumption at existing coal-fired power plants. The development status of these applications ranges from positive research findings in a particular area (such as water-based nanofluids with enhanced thermal conductivity that could reduce water use in a plant) to the actual use of nanotechnology-based products in other industries that could be tested for deployment in coal-fired power plants.

It appears that today, most of the potential applications are more toward the research end of the deployment spectrum, and that relatively few are actually being deployed. This situation probably reflects the current economic situation and the state of the nanotechnology field in general. Nanotechnology is a new field, and the economic and business environments today do not favor expenditures on unproven technologies without some government assistance. At the same time, government funding for nanotechnology *research* has been fairly robust over the past few years (in response to federal government policies aimed at ensuring U.S. global leadership in the development and application of nanotechnology). The results of that research (largely at universities and national laboratories) are being seen in the literature today.

Moving from research to deployment in today's environment will be facilitated with federal support. Many of the potential applications identified in this overview warrant further RD&D. Examples of these include, but are not limited to, the following:

- Development of nanostructured membranes to reduce mineral scaling;
- Deployment of nano-based filters to inhibit corrosion in recirculating systems and protect RO membranes from fouling;
- Further RD&D on nanotechnology-based membranes for CO_2 capture;
- Deployment of coatings to insulate and reduce heat loss throughout the plant;
- Deployment of enzymes to improve fuel efficiency;
- Further RD&D on the commercialization of nanofluids for more efficient heat transfer;
- Development of nanotube-based sensing skins to detect strain and impact damage;
- Further RD&D on nanotechnology-based sensors for detecting pollutants in water and air; and
- Development and deployment of large-scale storage systems using nanotechnology.

By supporting further RD&D for some subset of these applications, reductions in water consumption could result and lessons learned from the supported RD&D could be applied to future efforts. To take advantage of this situation, it is recommended that NETL pursue investigation into funding further research, development or deployment for one or more of the potential applications identified in this report. An approach for doing so could involve the following steps:

- NETL identifies a subset of potential applications for which RD&D support would be appropriate, on the basis of a to-be-determined set of criteria.

- For each application in this subset (those meeting the criteria), further information regarding nature and level of support needed, along with potential costs and benefits and other data deemed appropriate by NETL, is collected.

- Findings of the collected information are distilled for use as a decision-making tool regarding appropriate RD&D funding.

Such a process allows NETL to take steps that are based on the best currently available information, to move at a pace consistent with its objectives regarding water reduction goals in power plants, and to make a significant contribution to furthering the development and deployment of technology aimed at mitigating a critical national issue.

7 REFERENCES

Bell, T., 2009, "Is Downsizing to Atomic Scale One Way Forward? Nanotechnology for Energy," *The Bent of Tau Beta Pi*, Summer, available at http://www.tbp.org/pages/Publications/Bent/Features/Su09Bell.pdf. Accessed June 25, 2010.

Choi, U.S., 2008, "Nanofluids: A New Field of Scientific Research and Innovative Applications," *Heat Transfer Engineering*, 29(5):429–431.

EPA, 2010, "Primary National Ambient Air Quality Standard for Sulfur Dioxide; Final Rule," *Federal Register*, Vol. 75, No. 119, pp. 35519–35603, June 22.

EPRI (Electric Power Research Institute), 2007, Program on Technology Innovation: State of Knowledge Review of Nanostructured Coatings for Boiler Tube Applications, Palo Alto, CA.. 1014805

EPRI, 2008, "Program on Technology Innovation: Computational Modeling and Assessment of Nanocoatings for Ultra Supercritical Boilers," July 23, available at http://my.epri.com/portal/server.pt?space=CommunityPage&cached=true&parentname=ObjMgr&parentid=2&control=SetCommunity&CommunityID=404&RaiseDocID=000000000001016181&RaiseDocType=Abstract_id. Accessed June 21, 2010.

EPRI, 2010, "Boiler Life and Availability Improvement Program – Program 63, Program Overview," available at http://mydocs.epri.com/docs/Portfolio/PDF/2010_P063.pdf. Accessed June 25, 2010.

Evans, D., ed., 2009, "Nanotechnology-Enabled Sensing, Report of the National Nanotechnology Initiative Workshop," Arlington, VA, May 5–7, sponsored by National Science and Technology Council, Committee on Technology, Subcommittee on Nanoscience, Engineering and Technology.

Intertek, 2010, "Waterwall Deposit Measurement System," available at http://w3dev.intertek.com/servicesdetail.aspx?id=8571. Accessed June 21, 2010.

Kennedy, S., 2010, "The End of the Oil Change?" *Additive Communications*, available at http://www.plantservices.com/articles/2006/030.html. Accessed June 25, 2010.

Kim, J-K, and R. Smith, 2004, "Cooling System Design for Water and Wastewater Minimization," *Ind. Eng. Chem. Res.* 2004, *43*, 608–613.

Kotov, N., 2010, personal communication between Nicholas Kotov, Department of Chemical Engineering, Department of Materials Science and Engineering, Department of Biomedical Engineering, University of Michigan, Ann Arbor, Michigan and Deborah Elcock, Environmental Science Division, Argonne National Laboratory, June 19.

Li, Q., Y.W. Li, Z. Liu, R. Xie, and J.K. Shang, 2010, "Memory Antibacterial Effect from Photoelectron Transfer between Nanoparticles and Visible Light Photocatalyst," *J. Mater. Chem* 20(6), 1021–1216, Feb 14.

Lin, N.H., M. Kim, G.T. Lewis, and Y. Cohen, 2010, "Polymer Surface Nano-structuring of Reverse Osmosis Membranes for Fouling Resistance and Improved Flux Performance," *J Mater Chem* 20, 4642–4652.

Loh, K.J., T-C. Hou, J.P. Lynch, and N.A. Kotov, 2009, "Carbon Nanotube Sensing Skins for Spatial Strain and Impact Damage Identification," *J. Nondestruct Eval* 28:9–25

NETL (National Energy Technology Laboratory), 2006, "Reduction of Water Use in Wet FGD Systems," Project Fact Sheet, Nov., available at http://www.netl.doe.gov/publications/factsheets/project/Proj432.pdf. Accessed June 25, 2010.

NETL, 2008, "Nanofiltration Treatment Options for Thermoelectric Power Plant Water Treatment Demands," Project Fact Sheet, available at http://fossil.energy.gov/fred/factsheet.jsp?doc=6233&projtitle=Nanofiltration%20Treatment%20Options%20for%20Thermoelectric%20Power%20Plant%20Water%20Treatment%20Demands. Accessed June 25, 2010.

NETL 2010, "Power Plant Water Management — Nanofiltration Treatment Options for Thermoelectric Power Plant Water Treatment Demands," available at http://www.netl.doe.gov/technologies/coalpower/ewr/water/pp-mgmt/nanofiltration.html. Accessed June 25, 2010.

Nolan, P.S., 2000, "Flue Gas Desulfurization Technologies for Coal-Fired Power Plants," presented by Michael X. Jiang at the Coal-Tech 2000 International Conference Nov. 13–14, Jakarta, Indonesia.

PWMIS, 2010, "Produced Water Management Information System, Fact Sheet — Membrane Processes," developed by Argonne National Laboratory for National Energy Technology Laboratory, available at http://www.netl.doe.gov/technologies/pwmis/techdesc/membrane/index.html. Accessed June 28, 2010.

Routbort, J., 2010, personal communication between Jules Routbort, Energy Systems Division, Argonne National Laboratory, and Deborah Elcock, Environmental Science Division, Argonne National Laboratory, June 17.

Saha, A., H. Roy, and A.K. Shukla, 2010, "Investigation into the Probable Cause of Failure of Economizer Tube of a Thermal Power Plant," *J. Fail. Anal. and Prevent.* 10:187–190.

Savage, N. and M.S. Diallo, 2005, "Nanomaterials and water purification: Opportunities and challenges," *J. Nanopart. Res.* (7), 331–342.

Sawant, P.D., Y.M. Sabri, S.J. Ippolito, V. Bansal, and S.K. Bhargava, 2009, "In-depth Nano-scale Analysis of Complex Interactions of Hg with Gold Nanostructures Using AFM-based Power Spectrum Density Method," *Phys. Chem. Chem. Phys.* 11(14), 2374–2378, available at http://www.rsc.org/delivery/_ArticleLinking/DisplayArticleForFree.cfm?doi=b816592k&JournalCode=CP. Accessed June 21, 2010.

Science Daily, 2004, "Titania Nanotube Hydrogen Sensors Clean Themselves," March 29, available at http://www.sciencedaily.com/releases/2004/03/040325073047.htm. Accessed June 25, 2010.

Shelley, S., 2008, "Nanosensors: Evolution, not Revolution . . . Yet," *Chem. Eng. Prog.*, June 1, available at http://www.aiche.org/uploadedFiles/Nano/Publications/060808a.pdf. Accessed June 21, 2010.

Singh, D., E. Timofeeva, W. Yu, J. Routbort, D. France, D. Smith, and KJ.M. Lopex-Cepero, 2009, "An Investigation of Silicon Carbide-water Nanofluid for Heat Transfer Applications," *J. Appl. Phys.* 105, 064306

Smith, K., W. Booth, and S. Crevecoeur, 2008, "Evaluation of Wet FGD Technologies to Meet Requirements for Post CO_2 Removal of Flue Gas Streams," available at http://www.icac.com/files/members/MEGA_2008_Dravo_K.Smith.pdf. Accessed June 21, 2010.

Tomar, V., 2008, "Computer Aided Multi-scale Design of SiC-Si3N4 Nanoceramic Composites for High-temperature Structural Applications," Apr. 10, available at http://www.netl.doe.gov/publications/proceedings/08/ucr/abstracts/Tomar_Abstract.doc.pdf. Accessed June 25, 2010.

Tomar, V., 2010, "Nanocomposite Ceramics — What are Nanocomposite Ceramics?" AZONano, available at http://www.azonano.com/details.asp?ArticleId=2501#6-12. Accessed June 25, 2010.

Wang, L., W. Chen, D. Xu, B.S. Shim, Y. Zhu, F. Sun, L. Liu, C. Peng, Z. Jin, C. Xu, and N.A. Kotov, 2009, "Simple, Rapid, Sensitive, and Versatile SWNT-Paper Sensor for Environmental Toxin Detection Competitive with ELISA," *Nano Lett.* 9(12), 4147–4152.

Wu, C. "Using Light to Disinfect Water — New Light-activated Catalyst Keeps on Working Even After the Lights Go Out," *MIT Technology Review,* Jan. 27, available at http://www.technologyreview.com/printer_friendly_article.aspx?id=24415&channel=biomedicine&section=., or http://www.technologyreview.com/biomedicine/24415/page1/. Accessed June 21, 2010.

Yu, W., D.M. France, D.S. Smith, D. Singh, E.V. Timofeeva, and J.L. Routbort, 2009, "Heat Transfer to a Silicon Carbide/Water Nanofluid," *Int. J. Heat Mass Transfer* 52, 3606–3612.

Zhao, Y., S. Wang, and D. Li, 2009, "Study on the Coal Combustion and Desulfurization Catalyzed by Nano-additives," paper presented at the 3rd International Conference on Bioinformatics and Biomedical Engineering, June 11–13, Beijing, China.

www.ingramcontent.com/pod-product-compliance
Lightning Source LLC
Chambersburg PA
CBHW081805170526
45167CB00008B/3338